ENERGY SECTOR STANDARD
OF THE PEOPLE' S REPUBLIC OF CHINA

中华人民共和国能源行业标准

Code for Seismic Design of Hydropower Projects

水电工程防震抗震设计规范

NB 35057-2015

Chief Development Department: China Renewable Energy Engineering Institute

Approval Department: National Energy Administration of the People's Republic of China

Implementation Date: March 1, 2016

China Water & Power Press
中国水利水电出版社
Beijing 2024

图书在版编目（CIP）数据

水电工程防震抗震设计规范 = Code for Seismic Design of Hydropower Projects（NB 35057-2015） : 英文 / 国家能源局发布. -- 北京 : 中国水利水电出版社, 2024. 3. -- ISBN 978-7-5226-2592-8

Ⅰ. TV-65

中国国家版本馆CIP数据核字第2024GW3019号

ENERGY SECTOR STANDARD
OF THE PEOPLE'S REPUBLIC OF CHINA
中华人民共和国能源行业标准

Code for Seismic Design of Hydropower Projects

水电工程防震抗震设计规范

NB 35057-2015

（英文版）

Issued by National Energy Administration of the People's Republic of China

国家能源局　发布

Translation organized by China Renewable Energy Engineering Institute

水电水利规划设计总院　组织翻译

Published by China Water & Power Press

中国水利水电出版社　出版发行

Tel: (+ 86 10) 68545888　68545874

sales@mwr.gov.cn

Account name: China Water & Power Press

Address: No.1, Yuyuantan Nanlu, Haidian District, Beijing 100038, China

http: //www.waterpub.com.cn

中国水利水电出版社微机排版中心　排版

北京中献拓方科技发展有限公司　印刷

184mm×260mm　16 开本　3 印张　95 千字

2024 年 3 月第 1 版　2024 年 3 月第 1 次印刷

Price（定价）: ¥ 500.00

Introduction

This English version is one of the China's energy sector standard series in English. Its translation was organized by China Renewable Energy Engineering Institute authorized by National Energy Administration of the People's Republic of China in compliance with relevant procedures and stipulations. This English version was issued by National Energy Administration of the People's Republic of China in the Announcement [2023] No. 4 dated May 26, 2023.

This version was translated from the Chinese standard NB 35057-2015, *Code for Seismic Design of Hydropower Projects*, published by China Electric Power Press. The copyright is reserved by National Energy Administration of the People's Republic of China. In the event of any discrepancy in the implementation, the Chinese version shall prevail.

Many thanks go to the staff from relevant standard development organizations and those who have provided generous assistance in the translation and review process.

For further improvement of the English version, any comments and suggestions are welcome and should be addressed to:

China Renewable Energy Engineering Institute
No. 2 Beixiaojie, Liupukang, Xicheng District, Beijing 100120, China
Website: www.creei.cn

Translating organization:

China Renewable Energy Engineering Institute

HYDROCHINA Engineering Consulting Corporation Limited

Translating staff:

WU Mingxin SUN Xushu ZU Wei

Review panel members:

JIN Feng	Tsinghua University
LIU Xiaofen	POWERCHINA Zhongnan Engineering Corporation Limited
WANG Jinting	Tsinghua University
GUO Shengshan	China Institute of Water Resources and Hydropower Research
GUO Jie	POWERCHINA Beijing Engineering Corporation

Limited

LIU Qing POWERCHINA Northwest Engineering Corporation Limited

JIA Haibo POWERCHINA Kunming Engineering Corporation Limited

National Energy Administration of the People's Republic of China

翻译出版说明

本译本为国家能源局委托水电水利规划设计总院按照有关程序和规定，统一组织翻译的能源行业标准英文版系列译本之一。2023 年 5 月 26 日，国家能源局以 2023 年第 4 号公告予以公布。

本译本是根据中国电力出版社出版的《水电工程防震抗震设计规范》NB 35057—2015 翻译的，著作权归国家能源局所有。在使用过程中，如出现异议，以中文版为准。

本译本在翻译和审核过程中，本标准编制单位及编制组有关成员给予了积极协助。

为不断提高本译本的质量，欢迎使用者提出意见和建议，并反馈给水电水利规划设计总院。

地址：北京市西城区六铺炕北小街 2 号
邮编：100120
网址：www.creei.cn

本译本翻译单位：水电水利规划设计总院
中国水电工程顾问集团有限公司

本译本翻译人员：武明鑫　孙旭曙　祖　威

本译本审核人员：

金　峰　清华大学

刘小芬　中国电建集团中南勘测设计研究院有限公司

王进廷　清华大学

郭胜山　中国水利水电科学研究院

郭　洁　中国电建集团北京勘测设计研究院有限公司

柳　青　中国电建集团西北勘测设计研究院有限公司

贾海波　中国电建集团昆明勘测设计研究院有限公司

国家能源局

Announcement of National Energy Administration of the People's Republic of China [2015] No. 6

According to the requirements of Document GNJKJ [2009] No. 52, "Notice on Releasing the Energy Sector Standardization Administration Regulations (*tentative*) and detailed implementation rules issued by National Energy Administration of the People's Republic of China", 96 energy sector standards (NB) including *Specification of Shale Gas Reservoir Description* are issued by National Energy Administration of the People's Republic of China after due review and approval.

Attachment: Directory of Sector Standards

National Energy Administration of the People's Republic of China

October 27, 2015

Attachment:

Directory of Sector Standards

Serial number	Standard No.	Title	Replaced standard No.	Adopted international standard No.	Approval date	Implementation date
…						
30	NB 35057-2015	Code for Seismic Design of Hydropower Projects			2015-10-27	2016-03-01
…						

Foreword

According to the requirements of Document GNKJ [2011] No. 252 issued by National Energy Administration of the People's Republic of China, "Notice on Releasing the Development and Revision Plan of the Second Batch of Energy Sector Standards in 2011", and after extensive investigation and research, summarization of practical experience, consultation of relevant Chinese and foreign standards, and wide solicitation of opinions, the drafting group has prepared this code.

The main technical contents of this code include: general provisions, terms, basic requirements, site selection and site categories, seismic safety evaluation of project sites, seismic design criteria, project layout and hydraulic structures, foundations and slopes, hydraulic steel structures, electromechanical facilities, communication, site access, earthquake monitoring, and emergency management.

In this code, Articles 1.0.5, 3.0.1, 4.1.1, 13.2.1, 14.1.1, 14.3.1, and 14.4.1 in bold are all mandatory provisions and must be strictly implemented.

National Energy Administration of the People's Republic of China is in charge of the administration of this code and the explanation of its mandatory provisions. China Renewable Energy Engineering Institute has proposed this code and is responsible for its routine management. Energy Sector Standardization Technical Committee on Hydropower Investigation and Design is responsible for the explanation of specific technical contents. Comments and suggestions in the implementation of this code should be addressed to:

China Renewable Energy Engineering Institute
No. 2 Beixiaojie, Liupukang, Xicheng District, Beijing 100120, China

Chief development organizations:

China Renewable Energy Engineering Institute

HYDROCHINA Engineering Consulting Corporation Limited

Chief drafting staff:

ZHOU Jianping	LI Sheng	YANG Zeyan	PENG Tubiao
YUAN Jianxin	DANG Lincai	ZHAO Kun	WEI Zhiyuan
WANG Huiming	YANG Zhigang	GONG Jianxin	HUANG Xiaohui
JIA Juntian	WANG Runling	FAN Junxi	YAN Yongpu
LI Fuyun	FANG Hui	NIU Wenbin	WANG Jilin

DU Xiaokai

Review panel members:

WANG Baile	CHEN Houqun	ZHANG Chuhan	GAO Mengtan
XU Xiwei	JIN Feng	AI Yongping	LI Deyu
LYU Mingzhi	WANG Renkun	FAN Fuping	XU Jianqiang
CHEN Yongfu	XIAO Feng	LEI Hongjun	

Contents

1 General Provisions

1.0.1 This code is formulated to specify the objectives, principles, criteria, and basic requirements for the seismic design of hydropower projects, to strengthen the seismic design of hydropower projects, and to improve the ability to protect against and mitigate earthquake disasters of hydropower projects.

1.0.2 This code is applicable to the seismic design of new large and medium-sized hydropower projects.

1.0.3 Earthquake-resistant work of hydropower projects shall be conducted at all stages of river planning, project design, and project operation, adhering to the policy of being people oriented and the principle of combining protective measures with rescue efforts while putting stress on the former, to prevent dam-break and minimize the losses caused by earthquake disasters.

1.0.4 In the hydropower planning of rivers and the design of hydropower projects, a study of seismic risk identification and fortification shall be made. Seismic measures shall be proposed for specific projects, hydraulic structures, and equipment and facilities by analyzing the potential seismic damage and hazards, and seismic requirements shall be proposed for project construction and operation management.

1.0.5 The seismic design of hydropower projects shall meet the following requirements:

1 The seismic design shall be conducted for large and medium-sized hydropower projects as per this code.

2 In an area with a basic intensity of Ⅶ or above, for a large hydropower project, or a medium-sized hydropower project with a dam higher than 70 m and complex seismic geological conditions, seismic design shall be made, and a special report shall be prepared in the feasibility study stage.

3 In an area with a basic intensity of Ⅷ or above, for a hydropower project with a dam higher than 200 m or a reservoir capacity over 10 billion m^3, the seismic safety of dam construction and seismic measures shall be demonstrated.

4 After a strong earthquake, for an in-service hydropower project in the area with an intensity of Ⅶ or above, the post-earthquake safety shall be checked, a special safety appraisal or evaluation shall be conducted, reinforcement and repair measures shall be taken, and

seismic measures shall be improved if necessary.

1.0.6 In addition to this code, the seismic design of hydropower projects shall comply with other current relevant standards of China.

2 Terms

2.0.1 seismic design

design of specific engineering and non-engineering measures to reduce seismic damage to a project and avoid secondary disasters, such as reasonable avoidance, structural reinforcement, and post-earthquake emergency disposal

2.0.2 seismic fortification class

classification of seismic designs for various structures, facilities, and equipment according to the factors that might cause casualties, direct and indirect economic losses, social impact after the project is damaged by an earthquake, and its role in earthquake rescue and relief

2.0.3 seismic design criteria

scale to measure the seismic design requirements, which is determined by the seismic fortification class, seismic design intensity, or design ground motion parameters. Generally, seismic design criteria are characterized by a return period based on probability theory

2.0.4 basic intensity

seismic intensity under ordinary site conditions with a 10 % probability of exceedance in 50 years, namely, the seismic intensity indicated by GB 18306, *Seismic Ground Motion Parameters Zonation Map of China*

2.0.5 design intensity

seismic intensity determined on the basis of the basic intensity for the seismic design of a project

2.0.6 secondary disaster induced by earthquake

disaster caused by earthquake damage to structures, facilities, and natural environments, such as fires, explosions, plague, pollution from toxic or harmful substances, floods, debris flow, and landslides

2.0.7 ultimate seismic resistance of dam

ability of a dam to resist strong earthquakes, which is the value of the maximum earthquake action that the dam can resist under specified conditions, i.e., no dam-break or uncontrolled release of a reservoir

2.0.8 ground motion monitoring

observation to record the ground motion or dynamic responses of hydraulic structures during an earthquake

2.0.9 earthquake monitoring

monitoring of seismic activities in and around the reservoir

2.0.10 emergency plan for earthquake

emergency action plan for disaster prevention and rescue, which is prepared in advance and different from normal working procedures in order to minimize the loss of life and property due to earthquakes and secondary disasters

2.0.11 frequently occurred earthquake

earthquake action with a 63 % probability of exceedance in 50 years

2.0.12 maximum credible earthquake (MCE)

largest earthquake magnitude that could occur along the faults within a given area or under the current tectonic framework or predicted to occur according to the historical seismic records

2.0.13 emergency shelter

living service facilities with the function of emergency refuge, which are planned, designed, and constructed to respond to earthquakes and other emergencies and taken as safe places for emergency evacuation and temporary living

3 Basic Requirements

3.0.1 The hydropower project shall maintain its functions under frequently occurred earthquake, limit the damage to the repairable level under the design earthquake (DE), and prevent sudden collapse or disasters under the safety evaluation earthquake (SEE).

3.0.2 The return period of earthquakes for a hydropower project shall be determined according to the site conditions, project rank, and structure grade.

3.0.3 The seismic fortification class of a hydropower project shall be determined according to the importance of structures, equipment, and facilities, their possible consequences due to earthquake damage, their roles in earthquake rescue and relief, and the difficulty in their restoration and reconstruction.

3.0.4 The seismic fortification class for the main hydraulic structures and facilities of hydropower projects shall meet the following requirements:

1 Class A: those, whose seismic failure might cause a major disaster or a serious secondary disaster, require special design.

2 Class B: those, whose function allows no interruption or needs restoration without delay after an earthquake or whose seismic failure might cause major disasters, require enhanced seismic fortification.

3 Class C: those require seismic design based on a 10 % probability of exceedance in 50 years (site basic intensity), except for those defined in Items 1, 2 and 4 of this Article.

4 Class D: those with small scales, whose seismic failure would not cause serious consequences and secondary disasters, may have moderately lower design requirements under certain conditions.

3.0.5 Water-retaining structures in seismic fortification Class A shall adopt a two-level seismic design, and the other hydraulic structures shall adopt a one-level seismic design. The objective of the first-level seismic design is to tolerate local damage of structures which may be operable after repair under design earthquake. The objective of the second-level seismic design is to keep the dam stable without collapse under the SEE.

3.0.6 The seismic fortification classes of hydraulic structures shall be determined according to their importance and the basic intensity of the site, as defined in Table 3.0.6.

3.0.7 Seismic fortification Class B shall be assigned to the tower (or bank-tower) intake, water conveyance system, powerhouse, centralized control building, substation facilities, switchyards, outgoing line yards, and important traffic facilities of large hydropower projects. Seismic fortification Class C shall be assigned to the tower (or bank-tower) intake, water conveyance system, powerhouse, centralized control building, substation facilities, switchyards, outgoing line yards, and important traffic facilities of medium-sized hydropower projects. An upgrade or downgrade in the seismic fortification class shall be demonstrated.

Table 3.0.6 Seismic fortification classification

<table>
<tr><th>Seismic fortification class</th><th>Grade of structure</th><th>Site basic intensity</th></tr>
<tr><td>Class A</td><td>Water-retaining and important water release structures of Grade 1</td><td rowspan="2">≥ VI</td></tr>
<tr><td>Class B</td><td>Non-water-retaining structures of Grade 1 and water-retaining structures of Grade 2</td></tr>
<tr><td>Class C</td><td>Non-water-retaining structures of Grade 2 and structures of Grade 3</td><td rowspan="2">≥ VII</td></tr>
<tr><td>Class D</td><td>Structures of Grades 4 and 5</td></tr>
</table>

NOTE Important water release structures refer to those that might endanger the safety of Grade 1 water-retaining structures in case of failure.

3.0.8 The structure foundation shall be identical in the seismic fortification class with the corresponding structure and may upgrade in the seismic fortification class after demonstration.

3.0.9 The seismic fortification classes of slopes or landslides directly related to the main structures of the project shall be determined as follows:

1 When collapses or landslides seriously endanger the dam safety or might cause dam failure, the slope or landslide shall adopt the seismic fortification class of the dam.

2 When collapses or landslides endanger the safety of main structures, the slopes or landslide shall adopt the seismic fortification class of the structures affected or a lower seismic fortification class.

3 When collapses or landslides do not endanger the safety of main structures, seismic fortification Class C or D may be used for the slope or landslide after analysis.

3.0.10 The seismic fortification classes of hydraulic steel structures of discharge structures and the water conveyance and power generation system, the main electromechanical facilities, such as generator units, electromechanical equipment, electrical equipment and their systems, and the power system communication equipment shall be the same as that of their corresponding structures.

3.0.11 The structures in emergency shelters shall adopt seismic fortification Class B or above. The small-sized structures without serious failure consequences induced by earthquakes may degrade in the seismic fortification class accordingly.

3.0.12 The relationship between the basic intensity and the peak ground acceleration at the hydropower project site shall be determined according to Table 3.0.12.

Table 3.0.12 Relationship between basic intensity and peak ground acceleration at Class Ⅱ sites

Peak ground acceleration (a_{max}) at Class Ⅱ sites	$0.04g \le a_{max} < 0.09g$	$0.09g \le a_{max} < 0.19g$	$0.19g \le a_{max} < 0.38g$	$0.38g \le a_{max} < 0.75g$	$\ge 0.75g$
Basic intensity of project site	Ⅵ	Ⅶ	Ⅷ	Ⅸ	≥ Ⅹ

4 Site Selection and Site Categories

4.1 Site Selection

4.1.1 Regional tectonic stability, fault activity, and earthquake-induced geological disasters shall be studied, and avoidance measures shall be taken in dam site selection. Dam site selection must follow the principles below:

1 The dam site shall not be located in seismic regions with a basic intensity of Ⅸ or above.

2 The main structures of the project shall not be built on active faults or capable faults.

3 If the site is located in a high-risk area, the consequences of the earthquake damage to the project must be analyzed.

4.1.2 The site area of a hydropower project shall be classified and comprehensively evaluated in terms of factors such as the tectonic activity, regional tectonic stability, and risks of earthquake-induced geological disasters, as summarized in Table 4.1.2, based on the investigation and research of the regional geology and tectonics, project geology, and main physical and geological phenomena. The dam site and power plant site should be located in a favorable area or an ordinary area for earthquake resistance, and unfavorable areas should be avoided. A dangerous area is not allowed to be selected as the site without sufficient demonstration.

Table 4.1.2 Identification of site areas

Site area category	Fault activity	Hazard of earthquake-induced geological disaster	Regional tectonic stability
Favorable	No active fault within 25 km of the site; with a basic intensity of Ⅵ	Low	Good
Ordinary	No active fault within 5 km of the site; with a basic intensity of Ⅶ	Moderate	Fair good
Unfavorable	Active faults with lengths less than 10 km within 5 km of the site; seismogenic structures with magnitudes $M < 5$; with a basic intensity of Ⅷ	High	Fair poor
Hazardous	Active faults with lengths greater than 10 km within 5 km of the site; seismogenic structures with magnitudes $M \geq 5$; with a basic intensity of Ⅸ	Very high	Poor

4.2 Site Categories

4.2.1 The site soils after foundation excavation of a hydraulic structure should be classified according to the shear wave velocity of the soil layers or rocks, as shown in Table 4.2.1.

Table 4.2.1 Identification of site soil

Type	Shear wave velocity of soil layer v_{se} or shear wave velocity of rocks v_s (m/s)	Descriptions and features
Rock	$v_s > 800$	Hard, relatively hard, and sound rocks
Hard soil	$800 \geq v_s\ (v_{se}) > 500$	Fractured and relatively fractured or soft and relatively soft rocks; dense sandy gravels
Moderately hard soil	$500 \geq v_{se} > 250$	Moderately dense and slightly dense sandy gravels; dense coarse and medium sand; hard clay or silt
Moderately soft soil	$250 \geq v_{se} > 150$	Slightly dense gravels, coarse, medium, fine, and silty sand, ordinary clay and silt
Soft soil	$v_{se} \leq 150$	Mud, mucky soil, loose sandy soil, miscellaneous fill

NOTE In the case of multi-layer soil, equivalent shear wave velocity v_s of the soil beneath the foundation is calculated by $v_s = d_0 / \sum_{i=1}^{n}(d_i / v_{si})$, where d_0 is the overburden thickness (m); taken as the smaller value between the overburden thickness and 20 m; d_i is the thickness of the ith layer of soil (m); v_{si} is the shear wave velocity of the ith soil layer (m/s); and n is layer number of the soil.

4.2.2 Sites shall be classified according to the type of soil and overburden thickness, as shown in Table 4.2.2.

Table 4.2.2 Identification of site categories

Type	Overburden thickness d_{ov} (m)						
	0	$0 < d_{ov} \leq 3$	$3 < d_{ov} \leq 5$	$5 < d_{ov} \leq 15$	$15 < d_{ov} \leq 50$	$50 < d_{ov} \leq 80$	$d_{ov} > 80$
Rock site	I_0	–					
Hard site	–	I_1					

Table 4.2.2 *(continued)*

<table>
<tr><th rowspan="2">Type</th><th colspan="7">Overburden thickness d_{ov} (m)</th></tr>
<tr><th>0</th><th>$0 < d_{ov} \leq 3$</th><th>$3 < d_{ov} \leq 5$</th><th>$5 < d_{ov} \leq 15$</th><th>$15 < d_{ov} \leq 50$</th><th>$50 < d_{ov} \leq 80$</th><th>$d_{ov} > 80$</th></tr>
<tr><td>Moderately hard site</td><td>–</td><td colspan="2">I_1</td><td colspan="4">II</td></tr>
<tr><td>Moderately soft site</td><td></td><td>I_1</td><td colspan="3">II</td><td colspan="2">III</td></tr>
<tr><td>Soft site</td><td></td><td>I_1</td><td colspan="2">II</td><td colspan="2">III</td><td>IV</td></tr>
</table>

4.2.3 For complex geological conditions of the foundation or rock slope and weak discontinuities or weak rock layers existing in the project site, the stability conditions of the foundation and slope shall be identified, the safety under the design seismic action shall be checked, the potential hazards shall be estimated, and treatment measures shall be proposed.

5 Seismic Safety Evaluation of Project Sites

5.0.1 The seismic safety evaluation result approved by competent authorities at the state or province level shall be used as the basis for seismic design of a hydropower project.

5.0.2 The content and scope of seismic safety evaluation for a project site shall be classified according to the importance of the project and the complexity of the seismic geological conditions. The following requirements shall be met:

1 Level Ⅰ safety evaluation includes active fault identification, probabilistic analysis and deterministic analysis of seismic risk, determination of site ground motion parameters, reservoir earthquake prediction, and evaluation of earthquake-induced geological disasters. It is applicable to areas with a basic intensity of Ⅶ or above and large hydropower projects with dams higher than 150 m or storage capacities larger than 1 billion m^3.

2 Level Ⅱ safety evaluation includes active fault identification, probabilistic analysis of seismic risk, determination of site ground motion parameters, and evaluation of earthquake-induced geological disasters. It is applicable to large hydropower projects other than those required for Level Ⅰ safety evaluation.

3 Level Ⅲ safety evaluation includes the probabilistic analysis of the seismic hazard and the check of the peak ground acceleration. It is applicable to medium-sized hydropower projects.

5.0.3 The seismic safety evaluation levels of sites may be raised by one level for large and medium-sized hydropower projects with complex seismic geological conditions or active faults within 5 km of the dam site.

5.0.4 For large hydropower projects where active faults exist within 5 km of the dam site and strong earthquakes might occur, the maximum credible earthquake (MCE), relevant parameters of the scenario earthquake, and seismic response spectrum may be studied in addition to the Level Ⅰ seismic safety evaluation.

6 Seismic Design Criteria

6.0.1 The performance objectives corresponding to the seismic design criteria shall meet the following requirements:

1 Frequently occurred earthquake: hydraulic structures, electromechanical facilities, and hydraulic steel structures maintain good performances, and their durability or operational reliability are not affected.

2 Design earthquake (DE): the structures maintain good integrity, their local yielding or local damage is allowed, and they may continue to be used normally after emergency repair or appropriate maintenance; the hydraulic steel structures (gates) of water release structures can be opened in an emergency; the electromechanical facilities of the water conveyance and power generation system are not seriously damaged and can resume operation after rush repair or taking emergency measures; the hydraulic steel structures of navigation structures can resume operation after repair.

3 Safety evaluation earthquake (SEE): the dam can remain basically stable, without a dam break leading to disasters or chain reactions and without major secondary disasters endangering the public safety.

6.0.2 The seismic design criteria and design ground motion parameters for hydraulic structures shall meet the requirements in Table 6.0.2.

Table 6.0.2 Seismic design criteria of hydraulic structures

Seismic fortification class	Design earthquake	Safety evaluation earthquake	Remarks
A	$P_{100}=0.02$	$P_{100}=0.01$ (or MCE)	Applicable to Grade 1 water-retaining structures and important water release structures of Rank Ⅰ projects; safety check under the SEE is conducted only for dams
	$P_{50}=0.1$	$P_{100}=0.05$	Applicable to Grade 1 water-retaining structures of Rank Ⅱ projects; safety check under the SEE is conducted only for dams

Table 6.0.2 *(continued)*

Seismic fortification class	Design earthquake	Safety evaluation earthquake	Remarks
B	$P_{50} = 0.05$		Applicable to Grade 1 non-water-retaining structures
	$P_{50} = 0.1$		Applicable to Grade 2 water-retaining structures
C	$P_{50} = 0.1$		
D	$P_{50} = 0.1$		
Performance objective	Local damage is allowed but repairable	Remain basically stable without dam breaking	

NOTE $P_{100} = 0.02$ means a 2 % probability of exceedance in 100 years. Others are also explained in the same way.

6.0.3 Seismic fortification shall be conducted by using the ground motion parameters with a 2 % exceedance probability in 100 years reference period, 5 % exceedance probability in 50 years reference period, and 10 % exceedance probability in 50 years reference period for slopes or landslide masses of seismic fortification classes A, B, and C or D, respectively.

6.0.4 The seismic fortification criteria of dams shall be specially demonstrated for large projects with a basic seismic intensity of Ⅷ or above and a dam height of over 200 m or a total reservoir capacity of over 10 billion m^3 and a high dam site close to active faults (less than 5 km) with a potential seismic magnitude of Ⅶ or above.

6.0.5 The seismic fortification criteria of the main hydraulic steel structures and electromechanical facilities shall be the same as or stricter than those of water release, water conveyance and power generation, navigation, and other structures where the equipment and facilities are installed.

7 Project Layout and Hydraulic Structures

7.0.1 The basic data shall be collected and various impact factors shall be comprehensively analyzed for the project layout design. The project layout scheme shall be selected according to the functional requirements of the project, considering the topographical and geological conditions of the dam site, to facilitate construction and operation management and to effectively prevent and cope with floods, earthquakes, geological disasters, and secondary disaster risks.

7.0.2 The water release facilities for lowering the reservoir level or emptying the reservoir shall be studied for projects with dams that have a design seismic intensity of Ⅷ or above. The rate of lowering the reservoir level or emptying the reservoir shall not only meet the safety requirements of the dam itself under an earthquake shock but also prevent a man-made flood threatening the safety of downstream areas.

7.0.3 Attention shall be paid to preventing the risk of earthquake-induced geological disasters in the project layout and structural design. It is required to strengthen the investigation of geological hazards for natural slopes, landslide mass, dangerous rock masses, debris flow, and other physical and geological phenomena in and around the project area, to analyze their stability and possible impacts in case of earthquake, and to put forward effective countermeasures.

7.0.4 The main structures of a project shall be preferentially arranged on hard, intact, and thick bedrock and shall be far from areas with unfavorable geological conditions. The feasible seismic measures shall be proposed for the fault zone, shear zone, and weak rock mass in the overburden or rock foundation.

7.0.5 Seismic requirements shall be considered in the comparison and selection of project structure types. An appropriate type of dam is selected by analyzing the seismic capacity of the dam according to the topographical and geological conditions of the dam site, dam materials, and construction conditions. In areas with high seismic intensity and narrow river valleys, priority shall be given to the layout scheme of the underground headrace and power generation system, and tunnels shall be used as accesses to dams and plants.

7.0.6 Sudden changes of the shape or weakening of the stiffness at the parts with large seismic responses shall be avoided in the design of hydraulic structures. The seismic requirements shall be considered in the design of structures, and the construction specifications and operation and maintenance requirements shall be established from the perspective of the seismic resistance of structures.

7.0.7 The seismic calculation and analysis for main hydraulic structures shall include the sliding stability, stresses, permanent deformation, and seismic liquefaction characteristics under the design earthquake. In addition, the overall stability and safety of a dam shall also be analyzed and evaluated under the SEE for the water-retaining structures belonging to seismic fortification Class A.

7.0.8 For the project with a basic intensity of Ⅷ or above, whose dam height is greater than 200 m or whose reservoir capacity is over 10 billion m^3, the seismic failure mechanism, failure mode, and maximum shock resistance of the dam shall be studied, and the seismic safety of the dam shall be analyzed and evaluated comprehensively.

7.0.9 According to the seismic analysis results, engineering analogy, as well as the seismic damage investigation and analysis of similar projects, the seismic performance and seismic safety of structures shall be comprehensively evaluated in terms of the structure natural vibration characteristics, deformation damage, stresses, overall stability, and failure mode. Seismic measures shall be proposed for the weak parts or links that cannot effectively resist earthquakes.

8 Foundations and Slopes

8.0.1 The seismic fortification of the foundations and slopes of hydraulic structures shall prevent sliding failure, seepage failure, and harmful deformation that might endanger the safety of structures under seismic actions.

8.0.2 The following macroscopic seismic hazards or seismic effects on the foundations and slopes of hydraulic structures shall be studied:

1 Possible vibration damage to structures caused by strong ground motion.

2 Possible sliding failure of the foundation caused by strong ground motion.

3 Surface faulting, including the crack propagation of joints and other discontinuities.

4 Abnormal deformation of the local terrain, landform, and stratum caused by strong ground motion.

5 Landslides, collapses, flying rocks, and debris flows caused by earthquakes.

8.0.3 The possibility of liquefaction shall be assessed first for weak intercalated layers, sand layers, or clay layers in foundations. Corresponding seismic measures shall be taken for possible liquefied soil layers, sand layers, or weak clay layers, according to the structure types and the specific geological conditions.

8.0.4 For discontinuities such as faults, fracture zones, and interlayer dislocations in foundations, especially gently dipping intercalated mud layers and rock strata that may be argillized, the possibility of their failure and excessive deformation under the seismic action shall be studied, and corresponding seismic measures shall be proposed.

8.0.5 The excavation and comprehensive treatment scheme of engineering slopes shall be determined according to the location, engineering geological conditions, and failure mode.

8.0.6 For an engineering slope of seismic fortification Class A or B, multiple methods shall be used to study the ground motion responses of the slope, analyze the dynamic stability, dynamic deformation, dynamic stresses, and failure mode, and comprehensively evaluate the seismic safety of the slope.

8.0.7 According to ground motion response analysis, considering the experience of similar projects, the seismic safety of a slope or landslide mass shall be evaluated comprehensively, and seismic measures shall be proposed accordingly.

8.0.8 In addition to shotcrete-anchorage supports, a gentler slope, or wider berm shall be set for the excavated slope in the area with a high seismic intensity, and deep excavation and high fill shall be avoided. Protective and reinforcement measures shall be taken for unstable rock masses beyond the excavation line.

9 Hydraulic Steel Structures

9.1 General Requirements

9.1.1 The structural strength and rigidity of gates and hoists shall meet the requirements of the seismic design criteria. The seismic measures and installation requirements of hydraulic steel structures and facilities such as gates and hoists shall be clearly defined in the design and bidding documents.

9.1.2 The type selection and layout of hydraulic steel structures shall be checked according to the topographical and geological conditions and the layout of hydraulic structures, taking into account the possible impacts of an earthquake and earthquake-induced disasters on the foundations and slopes of structures and hydraulic steel structures.

9.1.3 For a high dam project with a basic intensity of Ⅷ or above, the safety of the ship lift or ship lock used for emergent flood discharge shall be studied as one of the emergency response plans to prevent dam failure.

9.2 Gates

9.2.1 The dogging mode, dogging device, and dogging position of the gate shall be safe and reliable to avoid locking failure or damage caused by ground motion.

9.2.2 The dogging beam under out-of-service conditions shall be fixed firmly to avoid displacement under the seismic action.

9.2.3 A cover plate shall be set at the top of the gate slot and storage slot according to the site conditions and shall be reliably fixed.

9.3 Hoists

9.3.1 The seismic actions and the seismic responses of structures shall be considered in the foundation design of the hoist, and the subgrade and foundation must be firm and reliable.

9.3.2 For a service gate for water release, the hoist shall adopt both local and remote control. For an emergency gate or bulkhead gate for water release, the fixed hoist shall adopt local control, and remote control should be set as well. Small hoists shall be able to operate manually.

9.3.3 For a quick-acting emergency gate for water conveyance, the hoist shall adopt both local and remote control. For an emergency gate for water conveyance, the fixed hoist shall adopt local control, and remote control should be set as well.

9.3.4 The gantry crane on the dam crest or tower crest shall be equipped with an anchoring device or a dogging device to prevent the gantry crane from overturning and unintended movement under seismic actions.

9.3.5 The winch hoist shall be equipped with a device to prevent the steel wire rope from jumping the groove. The hydraulic power pack and the control cabinet of the hydraulic hoist shall have sufficient strength, rigidity, and stability to avoid overturning under the seismic action.

10 Electromechanical Facilities

10.1 General Requirements

10.1.1 The structural strength of electromechanical facilities shall meet the requirements of the seismic design criteria. The design peak ground acceleration shall be specified in the type selection and design documents, bidding documents, and technical documents of procurement contracts of electromechanical equipment.

10.1.2 The seismic measures and the technical requirements for installation of electromechanical facilities shall be proposed, and electromechanical equipment and cabinets shall be fixed firmly with the foundation to avoid overturning or moving under the seismic actions.

10.2 Electrical Facilities

10.2.1 Seismic design shall be performed for the electrical facilities in the following cases, and the seismic calculation shall comply with the current national standard GB 50260, *Code for Seismic Design of Electrical Installations:*

1 Electrical facility with a voltage of 330 kV or above in the area with a design intensity of Ⅶ or above.

2 Electrical facility with a voltage of 220 kV or above in the area with a design intensity of Ⅷ or above.

3 Electrical facility installed indoors on the second floor or above or installed on a platform outdoors in an area with a design intensity of Ⅶ or above.

4 The main bus of a generator or a high-voltage outgoing GIL in an area with a design intensity of Ⅶ or above.

10.2.2 Electrical equipment shall be selected considering the design earthquake. When the electrical equipment cannot meet the seismic design criteria, vibration isolating and damping devices or other measures shall be provided.

10.2.3 For the layout of power facilities, such as step-up substations and outgoing line yards, topographical and geological conditions and earthquake hazards shall be considered, and the sites prone to secondary disasters induced by earthquake shall be avoided.

10.2.4 The layout of electrical facilities shall be determined through a techno-economic comparison, according to the design intensity, project-specific conditions, and other environmental conditions, taking into account the general layout, operation, and maintenance conditions of the electrical facilities. When the design intensity is Ⅸ, the layout of electrical facilities shall meet the following requirements:

1 For the power distribution equipment with a voltage of 110 kV or above, high type, half-high type, and double-layer indoor arrangements should not be used.

2 A suspended structure should be used for the power distribution equipment of a tubular busbar with a voltage of 110 kV or above.

3 The clearance between two pieces of main equipment and between main equipment and other equipment or facilities should be increased appropriately.

10.2.5 When the design seismic intensity is Ⅶ or above, the seismic design for the installation of electrical facilities shall meet the following requirements:

1 Flexible conductors should be used for the leads of electrical equipment and the connecting wires between equipment, and a margin in length shall be considered. When a hard busbar is used, a flexible conductor or expansion joint shall be used for transition.

2 Electrical equipment and electrical installations shall be installed firmly and reliably. The mounting bolts or welding strength of equipment and installations shall also meet the seismic requirements.

3 The design of wheels and tracks for a transformer should be avoided, and the transformer shall be fixed onto the foundation. The conservator and the pipes connecting the auxiliary equipment to the transformer shall meet the seismic requirements.

10.3 Station Service Power Supply System

10.3.1 The reliability of the station service power supply system shall comply with the current sector standard NB/T 35044, *Specification for Designing Service Power System for Hydropower Station.*

10.3.2 The station service power supply system shall provide an emergency power supply for flood discharge facilities with seismic fortification Class A or B. Whether the emergency power supply is provided for flood discharge facilities with seismic fortification Class C or D shall be analyzed and demonstrated according to the specific conditions.

10.3.3 The emergency power supply shall be directly connected to the power supply system of flood discharge facilities. The power supply and its power distribution equipment shall avoid the secondary disasters induced by earthquakes and shall be arranged as close to the water release structure as possible.

10.3.4 The emergency power supply configuration shall consider the electrical load characteristics, and that for the flood discharge shall meet the operating requirements of the water release equipment.

10.4 DC Power Supply System

10.4.1 The DC power supply system is used as the working power supply for the control and protection system of a hydropower station and as an emergency power supply in the hydropower station in case of an earthquake. A 220 V valve-regulated lead–acid battery pack with a redundant configuration shall be used for the DC power supply system to provide power for lighting in the plant in case of emergency, for emergency lighting and evacuation exit lighting in the hydropower station and emergency operation of the main equipment of the hydropower station.

10.4.2 A special battery room shall be set for the DC power supply system in the powerhouse, and the battery shall be installed on a bracket support with the gravity center lowered and the load of floor lightened. In an area with a design intensity of Ⅶ or above, the storage battery pack shall be firmly fixed.

10.4.3 The battery pack of 300 Ah or above used in the switching station DC power supply system shall be installed on a bracket.

11 Communication

11.1 General Requirements

11.1.1 Two or more independent communication channels must be set and form a circular or circuitous communication network in power line communication for large and medium-sized hydropower stations with an outgoing voltage of 330 kV or above. Different communication modes should be used for two independent channels.

11.1.2 Two or more independent communication channels shall be set and should form a circular or circuitous communication network in power line communication for medium-sized hydropower stations.

11.1.3 Two or more independent communication channels must be set and shall form a circular or circuitous communication network for the communication between large and medium-sized hydropower stations and the cascade (regional) centralized control center. Different communication modes should be used for two independent channels.

11.2 Satellite Communication

11.2.1 The cascade (regional) centralized control center shall be equipped with the satellite communication ground station.

11.2.2 The fixed-satellite communication shall be set as the second standby communication channel between large and medium-sized cascade hydropower stations and centralized control centers.

11.2.3 The capacity and scale of small satellite ground stations for a large or medium-sized hydropower station shall be determined by the capacity and scale of the hydropower plants, and the ground stations shall have the functions of transmitting voices, data, and images.

11.2.4 Fixed-satellite communication stations should be provided for the living quarters and administration areas of hydropower stations with a satellite communication ground station.

11.2.5 Mobile satellite phones shall be equipped as appropriate in the cascade (regional) centralized control center, control rooms of hydropower stations, living quarters, and administration areas.

11.3 Communication Power Supply

11.3.1 For large hydropower stations and medium-sized hydropower stations with an outgoing voltage of 330 kV or above, the communication power system shall be provided with the automatically switched, reliable double-circuit AC power supply and shall be equipped with an independent and reliable standby DC power supply.

11.3.2 A working power supply and standby DC power supply shall be set for the communication power supply of a medium-sized hydropower station.

11.4 Communication Network for Hydrological Telemetry and Forecasting System

11.4.1 The communication network for a hydrological telemetry and forecasting system shall have reliable communication channels. Two or more independent communication channels shall be equipped in important telemetry stations and should adopt different communication modes.

11.4.2 A satellite communication mode should be selected as the main communication mode for a hydrological telemetry and forecasting system.

12 Site Access

12.1 Types of Transportation for Site Access

12.1.1 The requirements for earthquake rescue and relief shall be considered in the selection of the transportation ways for site access, such as highway, railway, waterway, and aviation for hydropower projects.

12.1.2 Independent main and auxiliary accesses shall be built when highway transportation is selected for large hydropower projects. The feasibility of combined transportation types of highway and waterway as well as combined transportation types of highway, waterway, and aviation shall be studied.

12.1.3 For large hydropower projects, when the transportation condition is so inconvenient that auxiliary access is difficult, the feasibility of waterway passages in reservoir areas and emergency helipads shall be studied.

12.2 Accesses and Their Facilities

12.2.1 The seismic fortification class of dedicated accesses and facilities for hydropower projects shall be set according to the scale and importance of the main structures of hydropower projects, considering the restoration difficulty of accesses and their roles in rescue, relief, and resumption of production. The seismic design shall be conducted in accordance with the seismic design specifications of relevant sector standards.

12.2.2 The seismic fortification class of site accesses for a large hydropower station with dam height more than 200 m or reservoir storage capacity greater than 10 billion m^3 shall be set not inferior to Class B, and the seismic action shall be determined by the basic intensity of the site. The seismic measures for key sections or facilities shall be strengthened as required by one intensity level higher than the basic intensity of the site. After an earthquake of design intensity, the traffic works may be restored immediately after general rush repair.

12.2.3 Except for the large hydropower stations specified in Article 12.2.2 of this code, the seismic fortification class of site accesses for other large hydropower stations shall be Class C, and their seismic actions and seismic measures shall be determined by the basic intensity of the site. In case of an earthquake with the basic intensity of the site, the traffic works would not be collapsed or seriously damaged and can be restored immediately through short-time rush repair after an earthquake.

12.2.4 The seismic fortification class of site accesses for medium-sized hydropower stations shall be set not inferior to Class D. The seismic fortification shall be performed by comparing and contrasting the requirements of the basic intensity of the site, and the seismic measures may be appropriately lowered. When suffering an earthquake with basic intensity, the bridges, tunnels, and important structures of traffic works shall be free of serious damages.

12.3 Transport Routeways and Arrangement of Key Facilities

12.3.1 The locations of transport routeways, bridges, tunnels, wharves, and heliports shall be selected through an intensive investigation of seismic geological conditions to arrange the main routeways, bridges, wharves, and emergency helipads in areas as favorable for seismic resistance as possible.

12.3.2 The locations of transport routeways, bridges, and wharves shall be selected to avoid the zones prone to seismic geological hazards, especially zones where collapses, landslides, debris flows, or even liquefaction of the foundation soil may occur during the occurrence of an earthquake. Tunnels should be adopted for mountain highways or mountain railways passing through unfavorable geological zones, and the support shall be strengthened at the slope of the tunnel inlet and outlet.

12.3.3 For transport routeways, bridges, and wharves passing through unfavorable or hazardous areas, measures shall be taken to prevent seismic geological hazards and secondary disasters, or standby accesses shall be set. The following requirements shall be met.

1 The transport routeways or bridges should be arranged in the relatively narrow part of fractured zones when they must cross over the seismogenic structures.

2 The transport routeways or bridges should be arranged at the footwall of the fault when they are parallel to the seismogenic structures.

3 The transport routeways should avoid deep excavation and high fill, high abutments and piers, high retaining walls, and long-deep cuts.

4 When a route passes through a liquefiable soil layer and soft soil zone, necessary subgrade treatment measures shall be studied.

5 The prevention and control measures for geological hazards and slope treatment measures in key sections of routeways shall be strengthened.

6 Necessary maintenance should be performed for existing old bridges, roads, or ferry piers.

12.3.4 Reliable earthquake-resistant structure types shall be selected for important bridges and tunnels.

12.3.5 The location of a wharf in a reservoir area shall consider the effect of drawdown under earthquake conditions.

12.3.6 The location of an emergency helipad may be planned in combination with the operation management facilities of the hydropower station, and the ground traffic condition between the take-off and landing sites and the main rescue points shall be smooth. The types of helicopters and the number of helicopters taking off and landing at the same time shall be considered in site planning, and a clearance condition for the helicopters taking off and landing shall be guaranteed.

13 Earthquake Monitoring

13.1 Project Area

13.1.1 Representative positions in the large project shall be selected for strong earthquake monitoring and seismic response monitoring of the hydraulic structures according to the project scale, site seismic geology, and structural seismic response characteristics.

13.1.2 For Grade 1 dam on a site with a design seismic intensity of Ⅶ or above and for Grade 2 dam on a site with a design seismic intensity of Ⅷ or above, seismic monitoring arrays shall be installed to monitor the seismic response of hydraulic structures.

13.1.3 If any active fault exists in the project area, a high-precision leveling network, three-dimensional network, and short base lines should be set up to monitor the deformation of the fault.

13.1.4 A strong-motion seismograph shall be installed on the solid observation pier or in the observation room, which is less affected by the surrounding structures. The seismograph shall be provided with an independent emergency power supply that shall be well maintained and managed to ensure a smooth power supply during strong earthquakes.

13.2 Reservoir Area

13.2.1 For the reservoir with a dam higher than 100 m and a storage capacity greater than 0.5 billion m^3, the hazard analysis of reservoir-induced earthquakes shall be conducted. An earthquake monitoring network shall also be established to monitor earthquakes in the reservoir area.

13.2.2 The earthquake monitoring network shall be designed according to the seismogenic conditions and seismic intensity of the reservoir area. The reservoir earthquake monitoring network shall be put into operation at least one year prior to the impoundment.

13.2.3 The data recording and transmission system of the reservoir earthquake monitoring network shall operate reliably to facilitate real-time observation and analysis, accurate judgment, and timely understanding of the development of reservoir earthquakes.

14 Emergency Management

14.1 Earthquake Emergency Response Plan

14.1.1 In the design of seismic measures for hydropower projects, the risk analysis of seismic damage and secondary disasters shall be performed, and the earthquake emergency response plan and emergency management requirements for the projects shall be put forward for possible risks and hazards from the perspective of disaster prevention and mitigation.

14.1.2 Earthquake risks shall be analyzed based on the project design and construction in the principle of being scientific, systematic, and foresighted, to identify hazardous and harmful factors, determine the conditions and possibilities of disasters, analyze the damage and loss that might be caused, and formulate appropriate countermeasures accordingly.

14.1.3 The earthquake emergency response plan shall include the hazard analysis of earthquake damage and secondary disasters, organization structure and its responsibilities, prevention and early warnings, information reporting, emergency response, emergency preparedness, training, and drilling.

14.1.4 The earthquake emergency response plan shall include, but not be limited to, the prevention of the following risks: power outages of the whole plant, flooding of the powerhouse, dam overtopping, gate failure, dam breach, failure of the dam foundation or slope, increasing leakage through the dam foundation and dam body, power and communication interruptions, and secondary disasters induced by earthquake, including fires, floods, explosions, harmful gas release, oil leakages, landslides and collapses, debris flows, and dammed lakes.

14.2 Earthquake Emergency Response Organization

14.2.1 For a large or medium-sized hydropower project, an emergency response organization shall be established. An emergency management department shall be set up for flood control, hydropower station fire control, geological disaster prevention as well as earthquake emergency response.

14.2.2 The emergency response organization and personnel shall be incorporated into the staffing plan of the hydropower station. In the operation period, emergency rescue training and emergency response drilling shall be regularly organized and carried out for sudden earthquakes and the like.

14.3 Earthquake Emergency Supplies and Their Storage Requirements

14.3.1 For a large or medium-sized hydropower project, necessary

earthquake emergency supplies shall be reserved, and an emergency supply depot shall be built.

14.3.2 The earthquake emergency supplies shall be allocated according to the requirements of the earthquake emergency response plan for the project. The emergency equipment and supplies shall be categorized and listed in terms of supplies for command, lifesaving, rescue, living, and public use, and their maintenance and management shall be specified.

14.3.3 The storage of emergency response supplies shall meet the relevant requirements. The site selection and construction of depots shall facilitate safe storage, transport and docking of the supplies. The scale and standard of the depots shall be determined according to the number of people to be evacuated and the management requirements. Conspicuous signs shall be established.

14.4 Emergency Shelters and Evacuation Spots

14.4.1 For large and medium-sized hydropower projects, emergency shelters and evacuation spots shall be set in project area.

14.4.2 Emergency shelters and evacuation spots shall meet the requirements of safety and accessibility, and they shall be comprehensively planned in the principle of proximity, convenience, and practicality, considering the needs of the construction period and operating period to meet the emergency management requirements of natural disasters, such as earthquakes, and emergencies. According to the site conditions and objective needs, several emergency shelters and evacuation spots may be set in project area, operation management areas, and the living quarters. At the entrance and exit of each emergency shelters and evacuation spots, conspicuous signs shall be established.

14.4.3 Green lands, squares, stadiums, and indoor venues, gymnasiums, and stations meeting the seismic requirements may be used as temporary emergency shelters for earthquakes. Open spaces, green belts, and wide roads not affecting traffic may be used as temporary evacuation spots.

14.4.4 An underground powerhouse shall be provided with temporary emergency spots for earthquakes. The temporary emergency spots should not be more than 500 m from the centralized workplace of the employees, should be close to the emergency access and emergency exit, and shall be properly ventilated during an earthquake to avoid the asphyxiation hazard caused by floods, fires, and toxic gas leakages. Temporary emergency spots shall be equipped with adequate self-rescue equipment and emergency supplies.

14.4.5 Planning and design of emergency evacuation routes shall meet the following requirements:

1 At least two emergency evacuation access ways shall be provided in both the project area and living quarters. Conspicuous signs shall be set at the entrance and exit of the emergency evacuation access ways.

2 The reservoir and river course downstream intended for waterway evacuation routes shall be equipped with necessary equipment and facilities for water rescue.

3 For the project with complex topographical and geological conditions around the project area and living quarters, emergency helipads or helicopter rescue facilities shall be set up where roads and waterways are vulnerable to earthquake.

4 The design of emergency evacuation access ways shall be incorporated into the planning and design of on-site access and access roads for the project.

Explanation of Wording in This Code

1 Words used for different degrees of strictness are explained as follows in order to mark the differences in executing the requirements in this code.

1) Words denoting a very strict or mandatory requirement:

"Must" is used for affirmation; "must not" for negation.

2) Words denoting a strict requirement under normal conditions:

"Shall" is used for affirmation; "shall not" for negation.

3) Words denoting a permission of a slight choice or an indication of the most suitable choice when conditions permit:

"Should" is used for affirmation; "should not" for negation.

4) "May" is used to express the option available, sometimes with the conditional permit.

2 "Shall meet the requirements of..." or "shall comply with..." is used in this code to indicate that it is necessary to comply with the requirements stipulated in other relative standards and codes.

List of Quoted Standards

GB 18306, *Seismic Ground Motion Parameters Zonation Map of China*

GB 50260, *Code for Seismic Design of Electrical Installations*

NB/T 35044, *Specification for Designing Service Power System for Hydropower Station*